# TOO MUCH PLASTIC

## How Did This Happen and What Can Be Done?

Meryl Loonin

San Diego, CA

Printed in the United States

**For more information, contact:**
ReferencePoint Press, Inc.
PO Box 27779
San Diego, CA 92198
www.ReferencePointPress.com



LIBRARY OF CONGRESS CATALOGING-IN-PUBLICATION DATA

Author: Meryl Loonin
Title: Too Much Plastic: How Did This Happen and What Can Be Done?
Description: San Diego, CA : ReferencePoint Press, 2026.
Includes bibliographical references and index
Identifiers: LCCN 2025005341 (print) | ISBN 9781678212087 library binding | ISBN 9781678212094 ebook

For compete cataloging-in-publication data please go to www.loc.gov.

# CONTENTS

# Living in the Age of Plastic

From food wrappers and grocery bags to toothpaste tubes and shampoo bottles, cups and straws to electronics, athletic pants to Barbie dolls—and lifesaving devices like IV tubes and prosthetic limbs—modern life runs on plastic. In a time frame shorter than the average human lifespan, the world became entirely reliant on plastic. Supermarkets are filled with it, and online purchases are swaddled in it. Plastic makes fast food and fast fashion possible—and it is vital to the manufacture of cars, home appliances, and health care and hygiene products. Plastic is cheap to produce, lightweight, and sturdy. It is also remarkably versatile. With the right chemicals added to the mix during manufacture, plastic can be molded into almost any form imaginable.

## Threats to the Planet and Human Health

The same qualities that make plastic indispensable to modern life also make it an enormous threat to the planet. Plastic impacts the earth's environment at every stage of its life cycle. At the start, extracting the fossil fuels that supply the raw materials for plastic consumes massive amounts of energy and emits hazardous climate-warming gases. And at the end, when it is no longer useful, plastic stays in the environment to disrupt or destroy fragile ecosystems. After a plastic item is discarded, it can take hundreds of years—or even a thousand—to biodegrade. Sometimes the forces of

> "Without plastic we'd have no modern medicine or gadgets or wire insulation to keep our homes from burning down, but with plastic we've contaminated every corner of Earth."[1]
>
> —Matt Simon, science writer

wind, sunlight, and crashing waves break it into tiny pieces that persist on land, in rivers, and in oceans. In a widely cited 2017 study in the journal *Science Advances*, researchers estimated that more than three-quarters of the 9.1 billion tons (8.3 billion metric tons) of plastic that has ever been produced worldwide has been released into the environment as landfill or litter and remains there in some form. (For comparison, 1 billion elephants weigh about 8.3 billion tons [7.5 billion metric tons].) Traces of plastic have been discovered on the earth's highest peak, Mount Everest, and in its deepest ocean canyon, the Mariana Trench in the Pacific Ocean. Plastic sickens, injures, or kills tens of thousands of marine and land animals every year. "Without plastic we'd have no modern medicine or gadgets or wire insulation to keep our homes from burning down," says science writer Matt Simon, "but *with* plastic we've contaminated every corner of Earth."[1] Environmental advocates argue that the incredible benefits of plastic are now outweighed by the risks to the planet caused by producing too much of it.

Plastic also endangers human health. "It's piling up, leaching into our water, and poisoning our bodies,"[2] writes ProPublica climate reporter Lisa Song. The people who are hardest hit live in poor, marginalized communities next to heavily polluting plastic plants, waste facilities, and sites where plastic is dumped and burned. Yet no one is immune from the effects of plastic pollution. Even in more affluent communities, people breathe plastic dust in the air and ingest bits of plastic and its chemical additives in their food and water. While the health effects to humans are not yet fully understood, evidence is growing that plastics can alter the chemical and hormonal balance in the human body.

## Out of Sight, Out of Mind

Despite growing awareness of the dangers of plastic pollution, the world is producing—and discarding—more plastic than ever

*Plastic is cheap and versatile and can be found almost everywhere in modern life, including lining the shelves at the grocery store.*

before. "We just keep producing, producing, producing plastic,"[3] a United Nations Environment Programme (UNEP) official warns. UNEP researchers say that packaging and disposable items like cups, bottles, wrappers, and bags are among the biggest contributors to global plastic waste. After a trip to the grocery store or a takeout meal, people toss these items in the garbage—or in their curbside recycling bins. From there, it is out of sight, out of mind. Yet only about 9 percent of the plastics used worldwide have ever been recycled, according to the *Science Advances* study. The bulk of global plastic waste is buried in landfills, incinerated, or ends up as litter. In the environment, it accumulates on roadsides, clogs rivers, and flows into the ocean.

> **"We just keep producing, producing, producing plastic."[3]**
>
> **—United Nations Environment Programme official**

Humankind is leaving so much plastic in the environment that it often fuses with natural rock and sediment to become part of the earth's fossil record. Some researchers suggest that this signals a new geological era in the history of the planet: the plasticene epoch, or the plastic age. For years, environmental advocates say, individual consumers have been urged to change their behavior to reduce plastic waste. But no matter how many tote bags people carry to the store or water bottles they refill, they cannot avoid the plastics that flood the market. Rather than focusing on individual behavior, environmental activists in some of the most plastic-polluted parts of the developing world are charting a new way forward in the movement to curb plastic pollution. They are publicly calling out corporate plastic polluters, pressuring their governments for stricter regulations, educating their communities about the health risks of plastic, and building networks with other activists around the world. "I'm always trying to think, change the system, not change the people," says Tiza Mafira, founder of the nonprofit Plasticdiet Indonesia. "The system needs to change so that people will then find it easier to change their behavior."[4]

CHAPTER ONE

# The Plastics Revolution

New York–based journalist A.J. Jacobs thought he was setting a realistic goal for himself when he vowed to go plastics-free for a day. He figured his twenty-four-hour experiment would help him determine which plastic items were necessary to his life and which he could easily give up. But his experiment got off to a rocky start. When he woke up and swung his feet from the bed onto the floor, he realized that his bedroom carpet was made of nylon—a type of plastic. "I was roughly ten seconds into my experiment, and I had already committed a violation,"[5] he writes. On most mornings, Jacobs would have scrolled through his smartphone apps, but he had stashed his phone in a closet the night before since the device contains plastic. To wash up, he swapped out his regular toothpaste, toothbrush, shampoo, hairbrush, and liquid soap—all of which were made with plastic or packaged in plastic containers—for nonplastic alternatives that he had purchased at a specialty store.

His next hurdle was getting dressed. He had selected a pair of 100 percent cotton pants, but could not avoid the plastic in the waistband, label, zipper tape, and pocket lining. When it came time for breakfast, Jacobs found that his home kitchen was practically off-limits. Appliances such as the toaster, refrigerator, and microwave had plastic parts. Cold foods like milk, yogurt, and berries, along with pantry items like cereal and bread, were stored in plastic bags, jugs, and tubs. Jacobs carried cloth bags and glass containers to the grocery store and purchased cereal and nuts from the bulk bins. He bought fruit but realized that the bar-code stickers on the produce were plastic. Rather than using a plastic credit or debit card or his plastic-filled phone to pay for groceries, he brought a sack loaded with coins and counted them out at the register.

That evening, to avoid turning on a plastic-coated lamp—or typing on a plastic keyboard—Jacobs lit a candle and took notes about his day using only paper and an unpainted pencil, since the yellow paint on pencils contains plastic. Then he went to sleep on cotton sheets spread on a wooden floor, because the cotton in his mattress and pillow was blended with polyester-plastic fibers. "Admittedly living completely without plastic is probably an absurd idea,"[6] he concluded. By his own count, Jacobs had racked up 164 violations, instances when he had touched plastic over the course of his single-day experiment.

## What Is Plastic?

The word *plastic* means "malleable," or easily molded. True to the name, plastics can be shaped into almost anything. Like 98 percent of the plastics used today, the products that Jacobs struggled to avoid in his daily routine were derived from mining, pumping, and drilling for oil and gas. After crude oil and natural gas are extracted from the ground, they are treated with high heat and chemicals to refine them into gasoline to power cars—or fuels to produce heat and electricity. The process also releases chemical by-products that are the main ingredients in plastic.

To make plastic, the chemicals are heated again—a process called cracking—to break them apart into separate units and then subjected to tremendous pressure, until the units are thrust together into long, repeating chains of molecules. These ultra-long, unbroken chains of molecules are called polymers (Greek for "many parts"). They are synthetic, human-made substances, longer than polymers found in nature. And they are what gives plastic its most prized feature: its plasticity. Like a long beaded necklace that can be pulled, twisted, or coiled, manipulating the

Bathroom supplies are commonly packaged in plastic.

lengths and shapes of the polymer strands gives the plastic different properties like stretchiness, sturdiness, or transparency.

The unprocessed polymers that are the building blocks of all plastic products are referred to as resins. In their liquid state, plastic resins are fine-tuned by adding chemicals and dyes that make the finished products glossy, nonstick, fire resistant, or waterproof, or confer a host of other properties. When the resin cools and sets, it is chopped into oval-shaped beads the size of lentils. These plastic pellets—known as nurdles—are shipped to manufacturing companies around the world that melt, mold, and bake them into a wide range of products, including toys, auto parts, outdoor furniture, and packaging.

## The Invention of Synthetic Plastics

The world's first synthetic plastic, Bakelite, was a by-product of coal mining. It was named for its inventor, US chemist Leo Baekeland, who forged it in his home laboratory in 1907. Bakelite was marketed as "the material of a thousand uses."[7] In an era when US households were first becoming electrified, Bakelite was used to make light switches, outlets, radios, and telephones. It was also molded into costume jewelry, kitchenware, and even gums for false teeth.

In the 1930s scientists began experimenting with new, more flexible synthetic plastics that, unlike Bakelite, could be reshaped with heat and still retain their chemical structure. During World War II, many of these new plastics were rushed out of chemistry labs to be deployed in the real world. With war raging across continents, global supply chains were disrupted. Materials like steel, rubber, and silk grew scarce. The US government invested heavily in the country's budding plastics industry to spur the development of new products to supply the troops. Plastics were lightweight and durable; they substituted for steel, aluminum, rubber, and other materials in the manufacture of aircraft and tank parts, gun housings, parachutes, and even the razors and combs issued to soldiers. Many of the plastics still in use today—including nylon, acrylic, Teflon, and polystyrene foam (often referred to by its trade name Styrofoam, although it has other forms)—were invented in this era and mass-produced to support the war effort.

## A Miracle Material

After World War II ended in 1945, companies that had contracted with the US government to produce plastic for military use turned their attention to peacetime markets. Early plastics industry entrepreneur James W. Sullivan recalled that when the war ended, "virtually nothing was made of plastic, and anything could be."[8] (Sullivan's company became known for manufacturing the first pink plastic lawn flamingos.) Sullivan and his fellow industry pioneers benefited from new technologies like screw injection molding machines that shot liquid plastic into metal molds to fashion thousands of identical products.

> **"Virtually nothing was made of plastic, and anything could be."[8]**
>
> **—James W. Sullivan, early plastics industry entrepreneur**

Soon US factories were churning out a stream of new plastic household items. Historians explain that a growing American middle class had emerged from the hardships and scarcity of wartime. People were flush with optimism and buying power. Among the plastic consumer products that debuted in the postwar years were the squeeze bottle, rainbow-colored Tupperware food containers, disposable Bic pens and cigarette lighters, and toys such as Hula-Hoops, LEGO bricks, and Frisbee flying discs.

The arrival of mass-produced plastic products revolutionized life in the United States and eventually the world. Items made from plastic were more affordable than those made from scarcer materials like metal, ivory, tortoiseshell, glass, and leather. For the first time, even people of modest means could afford to become consumers. Plastic items were easy to wipe and wash clean, so hygiene improved. They kept food from spoiling, so it could be transported greater distances and stayed fresh longer in the pantry or fridge. Kitchen items like plastic containers to store leftover food and disposable diapers that required no laundering or folding reduced the time that women spent on household and child-rearing chores. Plastic toys were considered safe for young children, since they were shatterproof and splinterproof, and could be quickly wiped clean.

*After plastic resin cools and sets it is cut into small plastic pellets, called nurdles. These are shipped to manufacturing companies who use them to create a wide variety of different plastic products.*

## Thirsting for Plastic Soda and Water Bottles

For decades, people drank soda from glass bottles, which they returned to the store for a deposit. The glass bottles were then trucked to bottling plants for cleaning and refilling. In the 1960s beverage companies started selling soda in aluminum cans. And in the 1970s they introduced the plastic soda bottle. The lightweight, shatterproof plastic bottles not only were cheaper to produce than glass bottles, but also changed the world's drinking habits. It became more convenient for people to chug soda on the go.

In the 1990s soda companies landed on a new way to boost profits: bottled water. To persuade consumers to pay for water instead of drinking it from their taps, the companies sold a fantasy of water sourced from sparkling mountain streams. Yet a recent study found that 64 percent of bottled water sold in the United States is filtered tap water. And in taste tests, consumers often choose tap water over bottled water.

Around the world, the growing thirst for water and other drinks bottled in plastic takes a heavy toll on the environment. A United Nations report estimates that roughly 1 million plastic water bottles are sold every minute worldwide, with 85 percent of them likely to become trash or litter.

Plastic also changed entire industries. In the manufacture of cars and airplanes, plastic parts replaced steel and other metals, making vehicles lighter and more fuel efficient. And in hospitals, plastic was a lifesaver. Surgical gloves, syringes, blood bags, and IV tubes could be used and discarded, preventing the spread of disease and making once-expensive, time-consuming medical treatments accessible to more patients. "Plastic promised abundance on the cheap," says science and culture writer Susan Freinkel. "No wonder we became addicted to plastic."[9]

## The Birth of a Sprawling Plastics Industry

According to Freinkel and other plastics historians, the surge of new plastic consumer products in this era coincided with the rise of the petrochemical industry. During World War II fossil fuel and chemical companies helped propel the United States and its allies to victory by powering armored tanks, planes, and heavy machinery. After the war the federal government backed fossil fuel companies as they expanded to meet the growing demand for gas-powered cars, planes, and electricity. Since chemical companies relied on fossil fuel by-products to make their fertilizers, plastics, and other materials, the two industries eventually merged into a powerful petrochemical industry.

Today a short list of multinational corporations dominates the global market for plastic resin. Topping the list are the world's biggest petrochemical producers, including ExxonMobil, Shell, Chevron Phillips, and Dow Chemical. In recent years Chinese petrochemical companies have joined their ranks. These companies keep their oil and gas refineries operating twenty-four hours a day, generating the chemical by-products for plastic resins. Plastic is cheap to produce and hugely profitable.

In the United States plastic production has grown even more profitable with the recent boom in fracking (hydraulic fracturing). With fracking, fossil fuel companies drill for natural gas by injecting water, chemicals, and sand at extremely high pressure into cracks in the ground to release gas deep beneath the earth's surface. Fracking supplies energy to power the electric grid; it also produces an abundant supply of the chemical by-products used in plastic resin. Materials experts say that fracking creates such a glut of the raw materials for plastic that it drives producers to flood the market with cheap disposable packaging. Fracking has helped make plastics one of the largest US manufacturing industries.

Besides the petrochemical companies, the plastics industry also includes tens of thousands of other US and global companies along the supply chain. These include companies that mold plastic into toys, auto parts, electronic devices, and other products; those that package the products; and the brands that rely on enormous quantities of plastic to drive their sales, including some of the world's biggest and most iconic food and beverage companies, such as Coca-Cola, Nestlé, PepsiCo, and McDonald's.

## Plastics A to Z

Although the word *plastic* is often used in the singular, the plastics industry actually produces thousands of different plastics. A

*Almost all children's toys and games are made of plastic.*

sleek, shiny surfboard; stretchy leggings; sparkly glitter; and durable plumbing pipe are all concocted from plastic polymers, each of which shape-shifts when the materials are tweaked and different chemicals added to the mix. Manufacturers are not required to list the chemicals in their products, but a team of international scientists with the PlastChem Project has identified sixteen thousand different chemicals present in plastic products, more than a quarter of which the scientists tag as being a concern to human and environmental health.

The most common types of plastic by their shorthand names include PET (or PETE), the flexible, clear plastic molded into water bottles and clamshell containers or spun into polyester fiber, and sturdy HDPE, the most widely used plastic in the world. HDPE is turned into milk and juice jugs, detergent bottles, and toys, while its softer, more pliable cousin, LDPE, becomes cling wrap, bags, and cups. PP resists moisture and bacteria, making it a favorite for tubs of yogurt and dairy products. Polystyrene foam (Styrofoam) is used to insulate and ship food, eggs, and hot beverages or in

## The Life Cycle of a Plastic Bag

The plastic bag is a miracle of modern engineering. Made from HDPE plastic, it is waterproof and light as a feather, yet strong enough to hold more than a thousand times its weight. The plastic bag was invented by a Swedish company in 1976. By the early 1980s, major US supermarket chains had made the switch from paper to plastic bags. The bags were cheap to produce and distribute, so stores began handing them out freely—even double-bagging grocery items. Americans started bringing home hundreds of plastic bags a year.

Today environmental groups estimate annual plastic bag use worldwide in the hundreds of billions—or even a trillion. Plastic bags are designed to be short-lived. Most end up in the trash, or worse, as litter. They fly out of garbage cans, trucks, and landfills and clog storm drains and gutters. Or they tumble over beaches and drift into the ocean, where they are smashed to bits by the waves or swallowed by marine animals that mistake them for jellyfish. In recent years, many US cities, states, and countries have banned or imposed fees on plastic bag use. But the bags are still a visible and costly source of litter.

building insulation. And strong, weather-resistant PVC (or vinyl) is a practical choice for pipes, high-tech cables, and raincoats—although it contains highly toxic chemicals that make it hard to dispose of safely.

From fast food to fast fashion, many modern industries are heavily reliant on these types of plastic. The UNEP estimates that in today's fashion industry, 60 percent of the material spun into clothing and shoes is plastic. The modern toy industry was built on plastic. Environmental groups estimate that as much as 90 percent of the materials used to make children's toys and games is plastic.

## The New Throwaway Living

Plastic's invasion into modern industries like toys and fashion did not happen overnight. When plastic products were first introduced, many people viewed them as cheap, shoddy imitations of the real thing. Allison Cobb, author of *Plastic: An Autobiography*, explains that it took aggressive marketing and industry-sponsored education campaigns to persuade consumers that plastics promised a new, enticingly modern way of life. "The advertising industry really rose along with the plastic industry, convincing us that we're lacking things, and we need to buy something to fix what we're lacking,"[10] she says.

In America, members of the generation that came of age during the Great Depression of the 1930s and World War II in the 1940s, when resources were scarce, learned to scrimp and save. They gave their shoes a second life with new heels and soles, mended clothing, washed and reused glass bottles, wrapped loaves of bread in tea towels to preserve them, and repaired furniture and toys. For the plastics industry, this generation's thriftiness was an obstacle. "The plastics industry realized quickly that a product that lasts forever does not have a good profit,"[11] says Cobb. Plastics makers began to market disposability as the key to their products' appeal. Those born into the wartime generation had to be taught to throw things away. But it did not take long before their children and grandchildren embraced the new lifestyle made possible by plastic. By 1955 one of the country's most popular magazines, *Life*, featured a cover story celebrating the new "throwaway living."[12]

**"The plastics industry realized quickly that a product that lasts forever does not have a good profit."[11]**

**—Allison Cobb, author of *Plastic: An Autobiography***

Early in the plastics boom years, few people thought about the impact that throwaway living would have on the environment—or what would happen to all that plastic at the end of its useful life. Yet according to environmental groups, plastic waste remained a manageable problem until the 1970s. The turning point came with a steep rise in the production of single-use plastics. According to the UNEP, packaging and single-use items make up nearly 40 percent of all the plastic produced today. This includes the shrink-wrapping, clamshell containers, cups, bottles, and bags that accompany an online purchase, a to-go order, or a trip to the grocery store. Along with plastic-filled cigarette butts, these products are among the biggest contributors to global plastic waste.

**"We take natural substances created over millions of years, fashion them into products designed for a few minutes use, and then return them to the planet as litter that we've engineered to never go away."[13]**

**—Susan Freinkel, science and culture writer and author of *Plastic: A Toxic Love Story***

As soon as the item is unwrapped or consumed, the plastic is tossed away. "We take natural substances created over millions of years, fashion them into products designed for a few minutes use, and then return them to the planet as litter that we've engineered to never go away,"[13] writes Freinkel.

By the early 2000s, UNEP researchers say that the world had disposed of more plastic in a single decade than in the previous forty years combined. In the years since, production of single-use plastics has grown at a staggering clip. Plastic is convenient and cheap to produce. It does not dissolve, rust, or break down quickly. Yet these same qualities that make plastic so versatile—and that have transformed modern life—also make its use extremely difficult to control.

## CHAPTER TWO

# The Threat to the Planet, Animals, and Human Health

At the Brevard Zoo's Sea Turtle Healing Center in Melbourne, Florida, veterinarians rescue and rehabilitate dozens of injured, sick, and stranded sea turtles every year. The center is located on a barrier island off Florida's eastern seacoast, which separates the calm waters of the Indian River Lagoon from the Atlantic Ocean. From spring to early fall, thousands of sea turtles journey to the barrier island's protected beaches to lay their eggs in the sand. Just beyond the coast, among the beds of seagrass, juvenile sea turtles forage for food. In recent years vets at the Sea Turtle Healing Center have treated a growing number of young sea turtles who have been sickened after ingesting bits of plastic. The bits become entangled in the seagrass and algae that is their natural food source. In many cases, by the time the turtles arrive at the center for treatment, it is a life-threatening emergency.

In a recent blog entry, center coordinator Jess Patterson reported that three young sea turtles—a juvenile green turtle and two hawksbill hatchlings—had been brought to the facility for care after being found stranded on nearby beaches. When the green turtle, whom the vets named Belle, arrived at the center, they could see bits of plastic protruding from her body. The plastic pieces were blocking her digestive tract and preventing her from eating. During surgery, the vets removed fourteen bits of plastic from Belle's body. The

*A sea turtle tries to free itself from a piece of plastic fishing net. Sea turtles can also get sick and die from eating pieces of plastic that they mistake for food.*

jagged edges had caused internal damage, but Belle was making a slow recovery. Sadly, the baby hawksbills could not be saved. An autopsy revealed twenty-four tiny fragments of plastic in one turtle's body, and the other turtle had also swallowed multiple pieces of plastic. The vets were stunned. "It's not uncommon to find one or two pieces," Patterson said, "but to find this many was heartbreaking. And not just in one, but both hawksbills."[14]

## A Planet Choking on Plastic

Plastic waste accumulates along the shores of coastal communities and threatens the lives of sea turtles and other marine life when residents jettison bags, bottles, and other single-use items. Researchers consistently measure high concentrations of plastic litter in the shallow waters next to densely populated coastal cities worldwide. But plastic waste can also travel far from its source. Lightweight grocery and chip bags, food wrappers, straws, and bottles littered on land are carried by wind, rain, ocean currents, and seabirds for hundreds or even thousands of miles. Fragments of plastic that chip and break apart from larger items spill into storm drains and wastewater systems before flowing into rivers, lakes, and oceans. Other plastics are abandoned by commercial fishing fleets far out to sea and wash ashore with tides of trash onto the beaches of remote, unsheltered islands.

At some point everything made of plastic becomes waste. For disposable, single-use plastic items, the average usage time is measured in mere minutes or days; durable, longer-lived products may be used for years. The UNEP estimates that the world is producing roughly 440 million tons (399 million metric tons) of plastic waste each year—an average of 110 pounds (50 kg) for every single one of the roughly 8 billion people living on Earth. Americans more than make up their share, trashing more plastic per person than residents of any other country—an average of 286 pounds (130 kg) per person each year, according to a study in *Science Advances*. With plastic production soaring and expected to triple in the coming decades, a massive amount of plastic trash will linger on the planet in some form for centuries to come. "Our planet is choking on plastic,"[15] the UNEP warns.

## Getting Rid of Plastic Waste

Most Americans take it for granted that the plastic they toss in the trash will be picked up curbside from their homes and businesses and carted away. Or in rural communities, residents may haul their own garbage to transfer stations, where it gets picked up. In contrast, the United Nations estimates that 2.7 billion people around the world lack access to regular garbage collection. In many poor, under-resourced communities, plastic waste is scattered across the landscape, tossed in rivers, or dumped and burned in uncontrolled sites.

When it is properly discarded, the destination for most US plastic waste is a landfill. In a modern US landfill, garbage is buried in layers of dirt and clay and sealed in with thick foundations to prevent greenhouse gases from being released and toxic chemicals from leaching into groundwater. But before it can be buried,

## How Plastics Threaten a Soaring Seabird

Every fall the world's largest colony of Laysan albatross—more than 1 million birds—flock to remote Midway Atoll in the northern Hawaiian Islands to lay their eggs. The Laysan albatross is a large seabird, with a nearly 6-foot (1.8 m) wingspan, that soars long distances to hunt for food. When the albatross chicks hatch, they are flightless balls of fuzzy gray down. The parents take turns flying out to sea to scavenge for squid, fish eggs, and krill, which they swallow undigested and regurgitate into their chicks' mouths.

Since Midway Atoll sits at the edge of the Great Pacific Garbage Patch, a steady tide of plastic debris washes onto its shores. At sea, the adult birds scoop up plastic scraps, which they mistake for food. When they feed plastic to their chicks, it blocks their digestion and leaches toxic chemicals into their bodies. The plastic causes many of the chicks to starve to death. Marine biologists with the US Fish and Wildlife Service have autopsied hundreds of the dead albatross chicks. They find plastic bottle caps, cigarette lighters, straws, toys, toothbrushes, fishing line, and microplastics jangling around in their bellies. The birds' stomachs are filled with enough plastic to "stock the check-out-counter at a convenience store,'" one expert says.

Quoted in Susan Freinkel, *Plastic: A Toxic Love Story*. Boston: Houghton Mifflin Harcourt, 2011, p. 116.

plastic often escapes and becomes litter. Lightweight bags and other packaging materials are caught in the wind and fly out. Or they are carried off by seagulls, rats, and other animals scrounging for scraps of food. The plastic that remains in the landfill does not biodegrade, but with time and weather, small pieces break off from bigger items. The plastic pieces drop into the leachate, the noxious liquid sludge that seeps to the bottom of the landfill and gets pumped out to wastewater treatment plants.

When plastic waste is not landfilled, it is often burned. Plastic is one of the most flammable substances on earth. It is also laden with toxic chemicals. According to the US Environmental Protection Agency, just 12 percent of US plastic waste is incinerated in controlled facilities, where it is often converted to energy. In developing countries, it is more common to burn plastic waste—and it is often set aflame in aging, accident-prone incineration plants or open-air dump sites. Environmental groups say that even in a controlled US waste-to-energy facility, burning plastic pumps more climate-warming gases into the sky than a coal-fired power plant. Burning plastic also leaves behind an ash so hazardous that it must be isolated in holding ponds or landfills. When plastic is set ablaze in an uncontrolled site, toxic chemicals leach freely into air, soil, and water. Open-air waste burning is banned in many countries, but the laws are often poorly enforced, so plastic is burned illegally.

## Tiny Pieces of Plastic with a Big Footprint

While plastic does not naturally biodegrade, when it is exposed to the sun's ultraviolet rays or battered by wind, rain, and waves, it breaks apart into smaller and smaller pieces. By the time plastic slips into the liquid leachate at the bottom of a landfill—or is carried by wind and rain into the seagrass where young sea turtles feed—the pieces are so small that they no longer look or feel like the original article.

Small particles of plastic that splinter and shed from larger items are known as microplastics. Scientists define microplastics as pieces of plastic smaller than 5 millimeters—about the diameter of a pencil eraser. As a plastic item deconstructs and becomes brittle and worn, the toxic chemicals that were added during manufacture leak out, and the potent climate-warming gas methane is released into the air. Scientists warn that these tiny pieces of plastic have a big environmental footprint.

Microplastics usually start as larger plastic items that get junked and over time split into smaller pieces. But there are many other ways that plastic pieces can escape into the environment. With the wear and tear of synthetic rubber car tires, plastic particles spin away onto road shoulders. When it rains, they mix with the soil and are rinsed into storm drains that flow into rivers and lakes and onward to the ocean. When clothing spun from plastic-based fabrics—like stretchy jeans, leggings, and fleece—churns through a washing machine cycle, it sheds hundreds of thousands of plastic microfibers. The microfibers flow to wastewater treatment systems, where some are captured by filters but others escape into waterways.

Microplastics are also released via other routes. Many beauty and personal care brands add plastic microbeads to their cosmetic products, facial scrubs, and cleansers to buff and polish skin. (In the United States and Europe, laws prohibit the use of microbeads in some products, but many cosmetics are exempt.) When people rinse the products off their skin and down the drain, the beads flow into wastewater treatment systems, where some evade the filters. Nurdles often spill into the environment as they travel along the supply chain. And when smokers flick away the plastic-filled filters in the butts of their cigarettes and grind them underfoot, they too are releasing microplastics.

Plastic items break down, but they never disappear. When microplastics shrink to become specks so small they are no longer visible to the naked eye, they are called nanoplastics. Scientists have only recently become aware of nanoplastics and are studying the impact these teeny-tiny plastic bits are having on the planet.

## Plastic Soup in the Ocean

Oceans cover 71 percent of the earth's surface. With plastic use surging, it is inevitable that much of the world's plastic waste, including microplastics and nanoplastics, ends up in the ocean. Although scientists caution that it is difficult to measure plastic in the ocean with certainty, a Pew Charitable Trust report estimates that roughly 12 million tons (10.9 million metric tons) of plastic waste spill into the world's oceans every year. That is comparable to a garbage truck dumping a load of plastic into the ocean every 45 seconds of every day, 365 days a year.

In 1997 sailor and marine researcher Charles Moore brought ocean plastic pollution to the world's attention. While traveling from Hawaii to California, Moore and his crew veered off course to sail through the North Pacific Subtropical Gyre. The gyre is one of five major gyres in the world's oceans—enormous areas of circular currents powered by wind and forces created by the rotation of the earth. It is far from the nearest landmass, but as Moore sailed through it, he was shocked to see plastic debris swirling and bobbing at the surface. He called the area the Great Pacific Garbage Patch and traveled the world to share his story.

Based on Moore's early descriptions—and the name he coined for it—people envisioned a solid island of plastic trash. Instead, scientists say, the reality of ocean plastic is far more dangerous, and nearly impossible to clean up. The gyre acts like a gigantic whirlpool, sucking plastic waste into its center. Large plastic items sometimes swirl around at the surface, but much of the plastic looks like pieces of confetti—or flakes in a snow globe—drifting with the currents. When microorganisms, algae, and dead matter hitch a ride on the floating pieces, they sink to the ocean floor. Scientists explain that the fragmenting plastic pieces not only leach chemicals into the water but also act like a sponge, absorbing

*Plastic items floating in the ocean eventually fragment into tiny pieces known as microplastics. These fragments are highly toxic to marine life that ingests them.*

> "These waters are more akin to a soup to which humanity has added an unknown number of plastic items and pieces."[16]
>
> —Erica Cirino, science writer

new toxins from the seawater. This makes the pieces highly toxic to marine life that ingests them. There are trillions of these microplastics that spread through every level of the water column, from surface to seafloor. "These waters are more akin to a soup to which humanity has added an unknown number of plastic items and pieces,"[16] says science writer Erica Cirino.

## Animals on the Front Lines of the Plastic Waste Crisis

With so much plastic flowing through the oceans, marine animals are on the front lines of the plastic waste crisis. They are at risk of ingesting, suffocating on, or becoming entangled in plastic. Re-

### The Health Risks to Fence Line Communities

The people who bear the heaviest health burden for the world's surging plastic consumption live, work, and play in the shadows of petrochemical refineries and plastics manufacturing plants—or in the crowded slums and rural villages where plastic is dumped and burned. Known as fence line communities, these areas are usually poor, under-resourced, and home to historically marginalized racial and ethnic groups. Residents breathe tainted microplastic-filled air and are routinely exposed to toxic chemicals. They have elevated rates of cancer, birth defects, respiratory diseases, and other ailments. There are fence line communities in Malaysia, Indonesia, Vietnam, and many other developing countries.

In the United States the towns of St. James Parish along the Mississippi River in Louisiana are known by the grim nickname "Cancer Alley." According to the nonprofit Human Rights Watch, the area's poor Black residents have the country's highest risk of cancer from industrial air pollution. They are surrounded by more than 150 petrochemical plants, including many that make plastics. In 2024 a US court green-lit a massive new petrochemical complex with ten plastics plants, over the objections of a community-based organization that sued to stop it. A panel of United Nations human rights experts calls Cancer Alley a "form of environmental racism" that poses "serious and disproportionate threats" to residents' human rights.

United Nations, "Environmental Racism in Louisiana's 'Cancer Alley' Must End, Say UN Human Rights Experts," March 2, 2021. https://news.un.org.

searchers with the World Wildlife Fund suggest that almost every species group in the ocean has encountered plastic. Hundreds of marine species are known to be harmed by it. Sea turtles, dolphins, seals, and whales are caught in ghost nets, the industrial-sized plastic fishing nets that commercial fishing vessels lose or abandon due to rough weather or damage. Marine animals also become entangled in plastic packaging or get straws lodged in their noses and throats. When they ingest and fill their guts with plastic, animals are unable to eat and often starve to death.

Scientists explain that microplastics move up the food chain. Plankton and small fish ingest tiny plastic particles and get a dose of toxic chemicals that spread through their tissues. When a predator eats the small fish that ate the plastic, the toxins transfer to its tissues. "The more plastic soup increases in scale," writes environmental advocate Michiel Roscam Abbing, "the greater the number of animal species that can't avoid it in their menus."[17] Plastic has become one of the leading causes of death for some marine species and is even pushing a few, including hawksbill sea turtles, to the brink of extinction.

Sea turtles are vulnerable to the harms of plastic pollution at every stage of their lives. Adult turtles consume plastic bags and latex party balloons, mistaking them for jellyfish. Younger sea turtles graze for food closer to the coast, which is often heavily polluted with plastic. And sea turtle hatchlings get trapped in plastic that washes onto beaches, preventing them from reaching the surf. Gulls, ospreys, cormorants, and other seabirds are also tricked into thinking that plastic is food. As the birds hunt for fish near coastlines, they scoop up bottle caps, scraps of plastic bags and balloons, and even cigarette lighters in their beaks. Many seabirds also build their nests with scraps of plastic fishing net and packaging, which entrap and endanger their chicks.

**"The more plastic soup increases in scale, the greater the number of animal species that can't avoid it in their menus."[17]**

**—Michiel Roscam Abbing, environmental advocate**

Farther from the ocean, plastics have been found in the bodies of freshwater fish

*Land animals such as elephants are often harmed by plastic after encountering it in their natural habitats, or when they feed at dumpsites. These elephants in Sri Lanka search for food in this landfill.*

and birds and land animals such as elephants, hyenas, zebras, camels, goats, and cows. These animals feed in landfills and dump sites—or ingest flyaway bags and bottle caps that invade their natural habitat. "All across the world, nonhuman animals are suffering the consequences of humanity's plastic addiction far more quickly than we are learning how exactly they, and perhaps, also we are being harmed,"[18] writes Cirino.

## Plastics in the Human Body

Evidence is growing that plastic is making humans sick, too. As people move through a plastic-filled world, they inhale plastic particles in the air and ingest them when their food and drinks are wrapped and served in plastic. Plastic also touches the skin when people wear synthetic clothing, sit on plastic furniture, play with plastic toys, and use plastic-filled electronic devices.

Nearly everyone carries traces of microplastics in their bodies. Medical researchers have detected plastic particles in human blood, lung tissue, digestive tracts, heart muscle—and even in the placenta that feeds and nurtures a growing fetus. Much of the research has focused on the risks of exposure to toxic chemicals in plastics, including flame-retardants, which keep products from catching fire, and plasticizers used to make them soft and flexible. "If a plastic makes its way into us, it's carrying those chemicals with it,"[19] says plastics researcher Sherri Mason.

Some of the chemicals in these microplastics are linked to elevated cancer risk. Others disrupt the endocrine system—the glands and organs that release hormones and affect growth, reproduction, and metabolism (the body's chemical processes for converting nutrients to energy). Studies show that pregnant women, infants, and young children are most vulnerable to disruptions in the endocrine system and other harms caused by the toxic chemicals in microplastics. While young children's bodies and brains are still developing, they interact with plastics in their baby bottles, pacifiers, and toys.

Scientists have also found that microplastics and their chemical additives migrate into food from plastic packaging, such as Styrofoam meat trays, egg cartons, and the coating on the inside of canned foods. The chemical exposure rises when plastic food containers are heated. People also get a regular dose of microplastics in their drinking water. In a study sponsored by Orb Media, scientists who tested global water samples found that people ingest microplastics with every sip of their tap water; those who swig from the eleven leading brands of plastic water bottles tested consume double the dose of plastic than they would get from tap water.

**"The effects of microplastics on human health, we maybe know 1 percent of it. We still don't know 99 percent of what they might do, or actually what they might not do."[20]**

**—Joost Smit, Dutch medical researcher**

Researchers are wary about the health effects of plastic-filled living but are still figuring out what it means for younger generations who are growing up immersed in a world of plastics. "The effects of microplastics on human health, we maybe know 1 percent of it," says Dutch medical researcher Joost Smit. "We still don't know 99 percent of what they might do, or actually what they might not do."[20]

# The Reality of Plastics Recycling

By sheer volume of waste, Sims Sunset Park in Brooklyn, New York, is the largest materials recovery facility—or MRF (pronounced merf)—in North America. Glass, metal, paper, and plastic discards that are collected from most of the 3.3 million households in New York City are brought to Sims Sunset Park to be sorted, separated, and sold to companies that recycle them into new products. It is a modern, state-of the-art facility that stretches for 11 miles (17.7 km) on a pier jutting out into New York Bay. The Sunset Park MRF operates round the clock, six days a week, powered by high-tech machines and a staff of eighty-five people who work in three separate shifts.

Every day, the facility receives 1,000 tons (907 metric tons) of glass, metal, and plastics that New Yorkers leave in curbside bins outside their houses, apartment buildings, and public schools. Their recyclables arrive in city sanitation trucks and barges and are emptied onto heaping mounds of waste on the plant's tipping room floor. From there, bulldozers push the waste into a giant, high-ceilinged warehouse filled with multilevel conveyor belts, chutes, spinning magnetic discs, air jets, and optical scanners. By the end of this obstacle course, glass, metal, and plastic are sorted into separate streams, and the plastics are prepped for either landfilling or recycling.

New York City accepts only hard plastic containers in its curbside recycling program, along with mixed-material drink cartons. Soft plastics are excluded. But that does not stop residents from throwing plastic bags, candy wrappers, toys, and even Christmas lights into their bins. Workers are stationed along the conveyor belts to catch and remove these contaminants, including bags and filmy food wrappers that can jam the machines. As the materials travel along the belts, small plastics like straws, cutlery, and to-go drink lids are ejected and fall to a lower level, where they are collected for landfilling. Machines dismantle the cartons, but only the paper is salvaged—the plastic scraps are landfill-bound. The remaining plastics pass through optical scanners that sort them by content, shape, and color. Those with the highest resale value are compressed into bales. The nonrecyclable plastics are trucked to landfills or incinerators; the bales of high-value plastics are loaded onto trucks and railroad cars and transported to manufacturers that wash, shred, and melt them into nurdles to be sold to make new products.

## Beyond the Curbside Bin: What Happens to Plastics?

Even for a large high-tech facility like New York's Sims Sunset Park, sorting and separating plastics is an enormous challenge. New Yorkers—like most Americans—are often disconnected from the reality of recycling. "They just think that the recycling cart is a portal to another universe, and it's not,"[21] says Monica Boehringer, a Virginia-based recycling coordinator. In the waste industry, this is known as *wishcycling*—the act of throwing things in the recycling bin that people think are or should be recyclable, even when they are not. To add to the confusion, US cities vary widely in the types of plastics they accept for recycling. Nationwide, the US Environmental Protection Agency estimates that 25 percent of the items chucked in curbside bins is nonrecyclable. And many of the nonrecyclable items are plastic, which has one of the lowest recycling rates of any material collected. Yet Americans keep discarding more plastic in curbside bins—and the costs of collecting and processing it fall to municipal governments and taxpayers.

*Materials collected curbside for recycling are processed at the Sims Sunset Park Materials Recovery Facility in Brooklyn, New York.*

Waste industry experts say there are many factors tipping the scales against plastics recycling. During processing, small items like straws and cutlery—and stretchy films and bags—clog the machinery. People often toss plastic containers in their curbside bins without removing food residue. For the MRFs, it is not worth the effort to clean gunky peanut butter jars or greasy takeout food trays. Most challenging for the MRFs is the fact that there are so many different types of plastic, each with its own physical properties, chemicals, and dyes, and they cannot be processed together. To be recyclable, every bale of plastic must have a high degree of purity. The optical scanners at a plant like Sims Sunset Park—or newer artificial intelligence–empowered technologies—make sorting plastics easier, but they are too expensive for most MRFs.

And even if the MRFs invest in high-tech sorting machines, waste industry experts say the biggest obstacle to plastic recycling remains. Plastic degrades when it is melted down and recycled. There are few markets for recycled plastics because new plastic resin is higher quality; it is also cheaper to produce. "It's not whether we want to recycle something or not," says California recycling specialist Amy Harris, "it's about can we sell it."[22]

## Rethinking Plastic Recycling's Success Stories

Even if it is technically possible to recycle many plastics, it is not economical to sort and process them because they have so little resale value. There are two exceptions: PET (#1), used in water and soda bottles, and HDPE (#2), used in milk and juice jugs, detergent containers, and shampoo bottles. They are widely viewed as plastic recycling's greatest success stories. Many of the bottles and containers in Americans' home kitchens are made of PET and HDPE, and they are the most recycled plastics in the world. Even so, plastic industry trade groups say that the rate of recycling for both plastics is low. For PET bottles, the US recycling rate is just under 30 percent, and it has hardly budged for decades.

### Plant-Based Bioplastics: Buyer Beware

As concerns about single-use plastic recycling grow, many stores have begun selling bioplastic cups, cutlery, bags, bottles, and other products. Bioplastics are often made from plant-based materials, including cornstarch, sugarcane, and potato starch. They are marketed as biodegradable or compostable. But scientists say consumers looking for eco-friendly alternatives to oil and gas-based plastics should be wary. Even if a product is labeled biodegradable, it can take months to decompose. And that happens only under the right conditions. In a landfill, deprived of air, sunlight, and water, the process takes much longer.

Scientists who have tested bioplastic products find that they contain many of the same toxic chemicals as conventional plastics, which leach into the environment as they biodegrade. Materials that are compostable break down under certain managed conditions, leaving nutrients in the soil. But unless the label explicitly says so, bioplastics will not break down in a backyard compost bin. Instead, they must be sent to a high-heat commercial or industrial composting facility. When people wishcycle bioplastics into curbside bins, they contaminate other materials—a growing problem for materials recovery facilities. Ultimately, environmental groups warn, bioplastics are throwaway, single-use items that replace other throwaway single-use items and do little to curb plastic pollution.

The US Food and Drug Administration (FDA) maintains stringent rules about which recycled materials can be used in food-grade packaging. Even if an item started its life as a food container, the FDA prohibits most plastics from ever being turned back into food and drink packaging because they can leach toxic chemicals that are unfit for human consumption. PET bottles are more likely to be melted down and reused to hold drinks when they are collected at grocery stores or bottle redemption centers, where they stay cleaner. PET bottles that are collected curbside and mingle with other trash are usually downcycled—turned into lower-quality products such as clothing, carpets, or construction

*Some companies turn recycled PET bottles and other plastics into clothing and blankets. Researchers say these items shed hundreds of thousands of microplastic fibers in the wash cycle.*

materials that cannot be recycled again. HDPE deteriorates faster than PET and is more likely to be downcycled into nonfood bottles, plastic lumber, utility pipes, crates, and flowerpots.

In the 1990s several eco-conscious clothing brands began turning recycled PET bottles into fleece clothing and blankets. The companies were widely praised for keeping single-use plastics out of landfills. But decades later, the discovery of microplastics has dimmed enthusiasm for clothing fashioned from recycled plastic. Scientists have found that fleece made from recycled PET sheds more microfibers in the washing machine than clothing made from new plastics. Water bottles made from recycled plastics may also pose a greater danger to human health. After examining ninety-one different studies of food and drink containers, researchers at Brunel University of London found that plastic bottles made from recycled PET leaked more harmful, hormone-disrupting chemicals into the water than bottles made from new plastic.

## How Industry Misled the Public About Recycling

For years many Americans have dutifully tossed plastics into their curbside bins believing that they would be recycled. Yet recent reporting reveals that plastic industry officials have long expressed serious doubts about recycling. A team of reporters from National Public Radio (NPR) and PBS's *Frontline* launched a joint investigation into plastics recycling. They spent months digging into internal industry documents and interviewing former plastics officials. The reporters uncovered reports, memos, and studies by company engineers dating back to the 1970s, which cited major obstacles to recycling, calling it "costly and difficult" and the sorting of plastics "infeasible."[23] Despite these misgivings, NPR reporter Laura Sullivan explains, "promoting the idea of recycling was a way to sell more plastic."[24]

**"Promoting the idea of recycling was a way to sell more plastic."[24]**

**—Laura Sullivan, National Public Radio reporter**

In the late 1980s plastics companies were facing public backlash. Many cities and towns—and even the US Congress—were considering laws to restrict single-use plastics like Styrofoam cartons and grocery bags. In response, the NPR-*Frontline* team says, the plastics industry went on the offensive. Petrochemical companies, packagers, and brands like Coca-Cola, PepsiCo, and McDonald's banded together to form industry trade and lobbying groups. They fought initiatives to curb the use of single-use products, while at the same time heavily promoting plastics recycling. They rolled out multimillion-dollar advertising campaigns, funded feel-good community recycling projects, and distributed recycling curricula to schools.

**"If the public thinks the recycling is working, then they're not going to be as concerned about the environment."[25]**

**—Larry Thomas, former plastics industry official**

At the time, many US municipalities were already collecting glass, aluminum, and paper curbside for recycling; industry trade groups pushed them to add plastics. "If the public thinks the recycling is working, then they're not going to be as concerned about the environment,"[25] former industry official Larry Thomas told the NPR-*Frontline* team. The strategy paid off. Americans eagerly took part in the new curbside recycling programs. And for years, the focus on recycling muted calls for bans on single-use plastics.

Environmental advocates say the hype about recycling was just that: hype. A recent report by the environmental group Greenpeace found that only 5 to 6 percent of annual US consumer plastic waste was being recycled. Environmental groups accuse plastics companies of greenwashing, or misleading the public about recycling to make consumers feel less guilty about buying more plastic. And they compare Big Plastic's tactics to those once used by Big Tobacco to peddle cigarettes while covering up internal studies showing that smoking greatly increased the risks of cancer and lung disease. After a flurry of negative reporting—including the NPR-*Frontline* exposé—industry officials have doubled down on recycling, saying the problems can be overcome with new advanced recycling technologies. In a press release for a splashy new media campaign called Plastic Recycling Is Real, industry executive Matt Seaholm wrote, "Plastic recycling is very real, and it happens every single day across America."[26]

## Deciphering the Plastics Codes

One of the plastic industry's most successful efforts to sell consumers on recycling was the adoption of the labeling system known as Resin Identification Codes (RICs). Today the RICs are familiar around the world: the triangle of three chasing arrows with a number in the center that is stamped at the bottom of most household plastics. An industry trade group introduced the codes in 1988 and, according to the NPR-*Frontline* team, spent millions of dollars lobbying states and countries to adopt them. The RICs categorize plastics into the six main types used in most consumer goods and packaging: #1 refers to PET and

### Unrecyclable Single-Use Sachets

Sachets are miniature, single-serving plastic pouches. In wealthier countries, they often hold condiments like ketchup. But in the developing world, the packets contain shampoo, detergent, instant noodles, coffee, and other products. They are marketed by multinational consumer brands in Southeast Asia, India, and Africa, where many people cannot afford to buy full-size products. Before the sachets appeared, shops measured small portions of sugar, coffee, and other goods for sale to poor customers, who brought their own containers. Now, customers tear a packet from long strips that hang in shops and stalls. After a single use, they toss them away.

The pouches litter roadsides, collect in landfills, clog rivers, and flow into the ocean. They are reported to have killed elephants in Sri Lanka and shrimp off the coast of Indonesia. The sachets' small size—and multilayered plastic and foil—make them unrecyclable. In Southeast Asia, millions of waste pickers eke out a living scavenging in landfills for materials to resell. They tell environmental watchdog groups that they leave the sachets behind because they have no value. "Our Earth is saturated with these uncollectable, unrecyclable, contaminated, valueless little packets," says Sian Sutherland, cofounder of British antiplastics group A Plastic Planet. "Now more than ever before we have to sack the sachet."

Quoted in Rebecca Smithers, "UK Must Act to Stamp Out 'Curse' of Plastic Sachets, Say Campaigners," *The Guardian* (Manchester, UK), February 25, 2020. www.theguardian.com.

#2 to HDPE. The other numbers are for items such as squeezable bottles, yogurt containers, and to-go cartons. The catchall #7 applies to any plastics that do not fit the other categories, including those used in electronics, baby bottles, and eyeglasses.

According to industry officials, the RIC system was never meant as a promise that plastic items could be recycled. Instead, it was designed to aid waste processors in sorting plastics. Yet author Oliver Franklin-Wallis, who interviewed dozens of global waste industry workers for his book *Wasteland*, says "the waste industry abhors it [the code system]."[27] At a facility like Sims Sunset Park, thousands of plastic items speed by on the conveyor belts every minute. Workers do not have time to stop and check labels. And even if they inspected every item, the codes would be of little help. Many products with the same numerical code—such as #1 water bottles and #1 clamshell containers—cannot be recycled together.

For consumers, the RICs are misleading. Numerous studies have shown that the chasing arrow symbol—the universal symbol for recycling—gives people the impression that everything with a number is recyclable. "No matter what your garbage service provider is telling you, numbers 3, 4, 6 and 7 are not getting recycled," says Eve O. Schaub, who wrote a memoir chronicling her yearlong attempt to live without nonrecyclable consumer products. "Number 5 is a veeery dubious maybe,"[28] she adds.

## Exporting Plastic Waste Abroad

For years after the introduction of the RICs, waste management companies were overwhelmed with hard-to-recycle plastic waste. In their desperation to find somewhere to stash the excess plastic, they began exporting it by the shipload to China, paying Chinese waste handlers to take it off their hands. Environmental groups say it was the US waste industry's dirty secret. "What do you do if you're the waste industry, collecting thousands of tons of mixed, contaminated or multi-layered plastics that—largely unknown to the general public—are impossible or unprofitable to recycle?" asks Franklin-Wallis. "You make it somebody else's problem."[29]

> "What do you do if you're the waste industry, collecting thousands of tons of mixed, contaminated or multi-layered plastics that—largely unknown to the general public—are impossible or unprofitable to recycle? You make it somebody else's problem."[29]
>
> —Oliver Franklin-Wallis, author of *Wasteland*

In the 1990s China was experiencing an industrial boom. Accepting plastic waste from abroad gave Chinese manufacturers access to a higher-quality raw material than they could make at home. And as the then most populous country in the world, China had a huge supply of cheap labor. Human rights watchdog groups reported that hundreds of thousands of Chinese workers labored long hours in brutal conditions sorting, shredding, and melting foreign plastic waste to be recycled into shoes, clothing, phones, holiday decorations, toys, and trinkets. The factories often operated illegally and lacked basic safety equipment. Many of the recycled products were exported back to the United States, where consumers lapped up cheap plastic goods labeled "Made in China."

In 2018 China abruptly closed its borders to foreign waste. The country no longer needed raw materials from abroad. And toxic plastic pollution had become a growing national concern. Watchdog groups described Chinese factory towns thick with smoke—and rivers contaminated with wastewater. China's about-face sent shock waves through the US waste industry. According to a study in *Science Advances*, at the time of the ban, the United States was shipping China as much as 70 percent of its plastic waste. For Europe, it was closer to 85 percent. Plastics began piling up in storehouses and landfills.

With media attention focused on the crisis, environmental groups say it should have been a wake-up call. Instead, enterprising waste handlers in developing countries such as Thailand, Indonesia, Malaysia, the Philippines, and Turkey, who were looking for easy profits, began to accept plastic waste from abroad. These countries have lax environmental and worker protection laws and lack garbage collection systems. Watchdog groups have documented abuses

*Plastic waste clogs a canal in Bangladesh. Illegal dumping of plastic waste has been documented throughout South and Southeast Asia.*

across South and Southeast Asia: waste handlers illegally dumping toxic plastic waste near crowded neighborhoods or in waterways and burning plastics in the open air.

In 2019, 187 countries—the United States not among them—signed an amended treaty, adding plastics to an earlier agreement that restricts the export of hazardous waste from richer countries to poorer ones. The treaty gives governments the right to refuse foreign plastic waste even if in-country waste handlers have agreed to take it. In the years since, the environmental justice group Basel Action Network has documented many violations by Canada, Europe, and the United States. The group says millions of tons of plastic waste still flow from richer to poorer nations every year. United Nations human rights officials call the global plastic trade "a terrible environmental injustice."[30]

## Recycle Only as a Last Resort

Critics argue that the industry's deceptions about recycling led to the rise of the global plastics trade, endangering the health of people in developing nations. They accuse plastics companies of driving up the production of wasteful single-use packaging while promoting a system of recycling they knew was deeply flawed. Americans have long been taught to reduce, reuse, and recycle. But for too long, environmental advocates say, the focus has been almost entirely on recycling. They propose that it is time to adopt a new slogan: reduce, reuse—and recycle *only* as a last resort.

CHAPTER FOUR

# Curbing Plastic Pollution

Hawaii's sandy beaches, vibrant blue waters, and diverse wildlife make it one of the top vacation destinations in the world. But Hawaii also has a serious plastic waste problem. The Hawaiian Island chain is located near the vortex of the Great Pacific Garbage Patch. Swirling ocean currents, strong winds, and crashing waves dump plastic debris from around the world onto its shores. Hawaiian environmental groups sponsor regular cleanups of the most polluted island beaches. Their crews find bottle caps, toys, cigarette lighters, toothbrushes, and other household plastics, along with heavy nets and gear lost or abandoned at sea by commercial fishing fleets. Every day, the waves also spew out small pieces of plastic that have broken apart from larger objects. Hawaii's eastward-facing beaches are coated with flecks of multicolored plastic that mix with the grains of sand. The plastic bits are nearly impossible for the crews to clean up.

Hawaiian beaches are also strewn with single-use plastic items that are littered on land or spill out of overfilled garbage cans. An estimated 9 million visitors flock to the Hawaiian Islands each year, and they leave behind a staggering amount of plastic waste. Cleanup crews recover beach toys, empty sunscreen bottles, Styrofoam containers, water bottles, and other junk. With limited space in its landfills, Hawaii is forced to pay to ship much of this plastic waste off the islands for disposal.

In 2015 Hawaii became one of the first US states to pass a ban on single-use thin plastic bags. In 2019 the city council in the state capital, Honolulu, followed up with one of the strongest bans on single-use plastics in the nation. The Disposable Food Ware Ordinance—also known as Bill 40—phases out nearly all single-use plastic food service items on the island of Oahu, including plates, cups, cutlery, straws, and Styrofoam containers.

Bill 40 was a victory for Hawaii's environmental groups, which had long fought to protect island beaches. But when the ban took effect in 2022, carrying it out proved to be a challenge. Many local stores and restaurants simply substituted bioplastic alternatives for oil- and gas-based single-use plastics. Scientists warn that bioplastics are often no safer than conventional plastics when released into the environment. And most bioplastics require a high-heat industrial composting facility to break them down. Yet the island of Oahu had no such facility. "Honolulu's plastic reduction effort is eliminating fossil fuel waste," writes local journalist Clare Caulfield, "but also reflects the struggle to find a solution that's both eco-friendly and convenient."[31]

## Banning the Bag and Other Single-Use Plastics

Environmental advocates argue that the only meaningful way to prevent plastic from spilling into the world's oceans—and washing ashore on beaches like Hawaii's—is by drastically reducing the amount of plastic that is made. "If your bathtub is overflowing, you wouldn't immediately reach for a mop, you'd first turn off the tap,"[32] says Annie Leonard, founder of the Story of Stuff Project, which raises awareness about the harms of excessive consumerism. One way to turn off the tap is with plastic bans like Oahu's Bill 40. In recent years, many cities, states, and countries have moved to ban or charge fees for the use of plastic bags, bottles, and other packaging.

> **"If your bathtub is overflowing, you wouldn't immediately reach for a mop, you'd first turn off the tap."[32]**
>
> **—Annie Leonard, founder of the Story of Stuff Project**

With its 2021 Single-Use Plastics Directive, the European Union (EU) is taking bold

*In an effort to reduce plastic pollution, many cities, states, and countries have banned the use of plastic bags. Research shows that such bans are working.*

steps to stem the tide of plastic waste. The directive, which will be phased in through 2030, targets ten items that are responsible for the bulk of marine litter on Europe's coastlines. It outlaws the sale of thin plastic bags, cutlery, plates, straws, balloons, and cotton bud sticks (Q-tips)—and requires that they be produced from nonplastic, biodegradable materials such as bamboo or paper instead. Each EU member nation must pass its own laws to comply with the directive.

In other regions of the world, countries in Africa and Southeast Asia have been at the forefront of the movement to ban single-use plastics. In Kenya plastic bags were becoming a health hazard. They clogged storm drains, collecting stagnant water that attracted malaria-carrying mosquitoes. In 2017 the Kenyan government passed one of the toughest bag bans in the world, making the selling or importing of plastic bags punishable with steep fines and prison time. Despite the harsh penalties, plastic bags are smuggled into Kenya from neighboring countries.

In the United States the record on plastic bag bans has been mixed. Industry groups fiercely oppose the bans and spend millions of dollars lobbying to block their passage. "Bans don't work," the industry trade group American Recyclable Bag Alliance asserts, "they punish people into using alternatives that are worse for the environment."[33] By 2025 twelve US states had passed plastic bag bans. But in a setback to the movement, twelve other states had bowed to industry pressure and passed legislation to prevent bag bans from being introduced.

At the local level, bag bans have gained more traction. Hundreds of US cities and towns have enacted bans or fees on plastic bags. A report released jointly by three environmental nonprofits measured the effectiveness of plastic bag bans in two US states and three cities. The researchers found that the bans reduced the number of plastic bags used each year by roughly 6 billion. "The bottom line is that plastic bag bans work,"[34] says environmental advocate Faran Savitz.

> **"The bottom line is that plastic bag bans work."[34]**
>
> **—Faran Savitz, environmental advocate**

## Back to the Future with Refill and Reuse Systems

Along with calling for single-use plastic bans, environmental advocates urge businesses to adopt refill and reuse systems. Rather than taking materials from the earth and quickly trashing them, the concept—called a circular economy—is focused on reusing, repairing, and reducing waste and pushing companies to extend the lifespan of their products. It harkens back to an era before disposable plastics, when soda and beer companies collected their empty glass bottles to be refilled, and milk was delivered in bottles that families returned to be cleaned and reused.

In the United States and Europe, eco-friendly local businesses—and some larger chain stores and restaurants—have introduced refill and reuse options. As the largest coffee company in the world, Starbucks pioneered to-go coffee culture. Today the global coffee chain uses more than eight thousand plastic and plastic-lined paper

*In response to pressure to reduce its plastic waste, Starbucks is encouraging the use of reusable cups.*

cups per minute. Under public pressure to curb its plastic waste, Starbucks is allowing customers to bring their own cups and, in some stores, to order drinks in reusable cups that are cleaned and refilled for other customers.

Another business model offers reusable to-go meal containers for customers who subscribe to a service through a mobile app. When they finish the meal from participating restaurants, they return the containers to designated drop boxes. In the United States refill and reuse systems have been slow to catch on. Refilling takes extra time and effort, and some people worry that reusable containers are unsanitary. Environmental advocates hope that as more cities ban single-use plastics, US consumers will adapt, and health and safety codes will be strengthened.

In many Southeast Asian countries, refill and reuse systems are already having an impact. In Indonesia, the Ciliwung River, which flows through the bustling capital city Jakarta, is among the world's most plastic-polluted waterways. Environmental attorney Tiza Mafira grew up in Jakarta and recalls the food carts that once passed by her home, dishing out noodles and meatballs into reusable ceramic bowls. In those days Indonesians ate to-go meals packed in banana leaves and carried woven baskets to shop. Today Styrofoam cartons and plastic sachets have displaced traditional refill and reuse systems. Mafira founded the nonprofit Plasticdiet Indonesia, which works with businesses to introduce packaging that can be cleaned, refilled, and reused. She describes it as using local wisdom revived for a modern world. "Our main inspiration is how it used to be done before the tsunami of single-use everything,"[35] she says.

## Buying the Beverage, Borrowing the Container

Another kind of refill and reuse system, bottle deposit bills, have been around for decades. Instead of banning plastic, they give people a monetary incentive to return bottles and cans for recycling. Ten US states, as well as some European countries and Canadian provinces, have enacted bottle bills. With a bottle bill in place, a store selling a soft drink such as Coke pays a deposit per bottle to the Coca-Cola distributor. The customer buys the Coke with a deposit added to the price—usually five to fifteen cents. When the customer returns the empty bottle, the deposit is refunded. The store also gets its deposit back plus a handling fee from the distributor for dealing with the empties. One environmental group describes it as "buying the beverage, but borrowing the container."[36] The incentive for consumers—getting the deposit back—is what makes bottle bills work.

And bottle bills do work. They have been around long enough for their impact to be measured. Bottle bills have been shown to reduce litter and keep plastic out of landfills. According to the Container Recycling Institute, bottle bill states with at least a ten-cent deposit fee experience an average return rate of 70 to 80 percent. In states without bottle bills, the return rate hovers closer to 20 to 30 percent. Bottle bills also supply a clean stream of PET plastic for recycling. Because the bottles are collected separately from other recyclables, they are less contaminated than bottles picked up curbside that travel through MRFs with a jumble of materials.

Recycling companies depend on the clean supply of PET from bottle bill states.

Despite this successful track record, beverage and packaging companies fight to kill bottle bills. The biggest beverage companies in the world—Coca-Cola, PepsiCo, and Anheuser-Busch—argue that reusing bottles threatens public health and hurts local businesses. Environmental groups counter that the companies' real concern is that deposits will raise drink prices, driving down sales. Coca-Cola—which produces more PET bottles than any other company—explains its opposition to bottle bills with an often-heard industry refrain: curbside recycling works better. Coca-Cola calls curbside recycling "one of the most effective and convenient recycling solutions for consumers."[37] The companies' lobbying efforts have been hugely influential in quashing bottle bills across the United States.

## Holding the Plastics Industry Accountable

For decades plastics companies fought to defeat bag bans and bottle bills while at the same time producing more single-use plastics. Environmental advocates accuse companies of introducing new plastic and mixed-material products without considering whether local waste systems can handle them. Municipal waste operators are overwhelmed and drowning in plastic. And taxpayers are left to foot the bill.

While high-profile companies like Amazon, Apple, and McDonald's have made voluntary pledges to reduce plastic waste, environmental watchdog groups say that companies often make such pledges only to backtrack when the heat is off. By contrast, extended producer responsibility (EPR) policies—also known as "polluter pays"—compel corporate plastic polluters through laws, regulations, or court orders to take responsibility for the life cycle of their products. The Story of Stuff Project's Annie Leonard explains EPR this way: "If you, Mr. Ketchup Producer, switch from a recyclable glass bottle to a squeezable one made of multiple plastic resins bonded together that can never be separated for recycling, you need to figure out how to deal with that at the end of its life."[38]

The environmental advocacy group Beyond Plastics has developed ten pillars to guide lawmakers in designing effective EPR policies. They include enforcing environmental standards for plastics packaging in the same way that fuel efficiency standards are required for new US cars, banning toxic chemicals in single-use plastics, and charging companies to recycle their plastic packaging. Some countries have passed EPR laws mandating that companies pay to manage plastic waste. In the United States EPR policies are gaining ground at the state level. Industry groups have tried to fend off passage of stringent state EPR laws by lobbying for bills that rely on voluntary compliance instead. "The EPR for

## Beach Cleanups and Brand Audits

Every fall, during the Ocean Conservancy's International Coastal Cleanup, hundreds of thousands of volunteers mobilize to pick up trash from beaches worldwide. For decades, beach cleanups were billed as feel-good, community-building events that helped protect the earth. Plastics industry groups sponsored their own cleanups, which they promoted with slogans like "People start pollution. People can stop it." Today environmental groups view the cleanups differently. They provide critical data on the sources of plastic pollution. In a recent cleanup, Ocean Conservancy volunteers collected 14 million pieces of trash. The top ten items found contained plastic, including cigarette butts, bottles, bottle caps, food wrappers, and bags.

The organization Break Free from Plastic raises the stakes of its global cleanups by holding corporate plastic polluters accountable. Volunteers are trained to conduct brand audits. They separate trash with visible labels and record the brand data. In a *Science Advances* study, researchers examined Break Free data collected in eighty-four countries. They found that just five corporations were responsible for more than a quarter of global plastic litter, chief among them Coca-Cola, PepsiCo, and Nestlé. "Brand audits transform beach cleanups into something truly powerful—a way to stop plastic pollution at the source by holding corporate polluters accountable," says Greenpeace's Graham Forbes, a Break Free from Plastic member.

Quoted in Oliver Franklin-Wallis, *Wasteland*. London: Simon & Schuster, 2023, p. 63.

Quoted in Jed Alegado, "Break Free from Plastic Conducts Massive Global 'Brand Audit' Actions," Break Free from Plastic, September 21, 2019. www.breakfreefromplastic.org.

> **"Can you imagine asking the tobacco industry to solve the smoking problem? Or asking fossil fuel companies to solve climate change? It simply won't work."[39]**
>
> **—Beyond Plastics**

packaging bills backed by industry put the polluters in the driver's seat and ask them to self-regulate," says Beyond Plastics. "Can you imagine asking the tobacco industry to solve the smoking problem? Or asking fossil fuel companies to solve climate change? It simply won't work."[39]

## Redesigning Plastics Packaging

While the world will not be able to innovate its way out of the plastic crisis, environmental groups view the redesign of plastic packaging as part of a broader effort to safeguard the planet and human health. Sometimes all it takes is a little tinkering with an existing product. For example, bottle caps are a major source of ocean pollution. As part of the EU's Single-Use Plastics Directive, beverage companies are required to redesign their bottles so the caps do not detach. To comply with the directive, Coca-Cola rolled out soda bottles with stay-put caps in several European cities. Some consumers have taken to social media complaining that the caps hit their noses and mouths when they drink. But EU officials urge people to give the new, more sustainable products a chance.

Other eco-conscious entrepreneurs see sustainable packaging as an opportunity to reap profits while also helping the planet. California-based firm Apeel has invented an edible, tasteless, and invisible plant-based spray for fruits and vegetables. The spray boosts the shelf life of produce like cucumbers and broccoli without the need for plastic shrink-wrap. Other packaging ideas are inspired by naturally occurring substances like seaweed, algae, and mushrooms. These are less resource intensive to grow and harvest than corn and sugar, which are used in many bioplastics. Notpla, a London-based company, uses seaweed to make ketchup packets, linings for cardboard containers, and ice cream spoons that melt away. The company says its seaweed-based products are edible or, if eating the package is not appealing, compostable in a backyard bin.

Environmental groups applaud the ingenuity of these products but say it is unlikely they can be scaled up to compete against fossil fuel–based plastics. They also warn consumers to beware of corporate greenwashing. Products touted as biodegradable, compostable, or ocean-friendly may not be what they seem. Materials experts worry that the new products are being developed—much like plastics decades ago—without first figuring out how they fit into existing waste systems.

## Holding On to Hope for a Global Solution

From bag bans to packaging redesign, efforts to curb plastic pollution often take hold at the local, state, or national level. Yet plastic waste does not stop at national borders. It flows through the world's rivers and oceans to wash up on distant shores. Or it is

### Zero-Waste Stores

In recent years, zero-waste stores have popped up in many US cities and towns. The goal of zero waste is to reduce the stuff people consume and relieve the planet of waste. Customers bring their own containers to shop for food and personal care and cleaning products, which are sold in bulk from bins, canisters, and liquid pump stations. Prices are determined by weight, so there is no need for wasteful single-use plastic packaging. The stores feature sustainable brands or goods from local businesses. Yet compared to big supermarkets, they are often expensive. And for shoppers accustomed to product-filled aisles, they offer limited choices. Zero-waste stores have caught on mainly in affluent US communities, where people have time and money to afford them.

In contrast, in the Philippines small Wala Usik (nothing wasted) stores are gaining a foothold in several cities and coastal towns. The Philippines is one of the world's worst marine plastic polluters. For decades, Filipinos bought small portions of food and household goods from family-run shops and food stalls. Today these goods are sold in plastic sachets. The Wala Usik stores replace sachets with micro-refilling stations for coffee, oil, sugar, and other goods. For Filipinos, they are a return to an older, more sustainable way of life.

exported as waste from richer nations to poorer ones, endangering the lives of residents in fence line communities. Along the way, plastic leaches toxic chemicals and greenhouse gases that threaten ecosystems and warm the planet.

With the stakes growing higher, in 2022 United Nations member nations agreed to come together to forge a global treaty to address the plastic pollution crisis. Over two years, delegations from 175 countries met in five different cities worldwide. During the final negotiating session in 2024, hopes of a treaty collapsed when a small group of petrochemical-producing countries led by Saudi Arabia, Iran, and Russia blocked attempts to impose limits on plastics production. The treaty talks were put on hold—with member nations agreeing to try again the following year.

As a youth ambassador for the Plastic Pollution Coalition and founder of the nonprofit River Warrior Indonesia, Aeshnina Aqilani was invited to attend the treaty talks as an observer. The river that flows near her home in East Java, Indonesia, is

*A forklift transports compressed and packaged bales of plastic bottles for export. Plastics that are sent to developing countries for recycling often end up in illegal dumpsites, incinerators, or waterways.*

polluted with sachets, Styrofoam, plastic bottles, diapers, and bags—much of it exported from Western countries. Addressing delegates after the failed global treaty talks, Aqilani said, "Your decision here to extend the process gives us hope that you will still deliver the treaty that the world and my generation need. . . . We urge you to listen to us, hear our voices, and most importantly do the right thing."[40]

Meanwhile, with the global treaty talks in limbo, Aqilani and her fellow environmental activists have vowed to press on with their work. Plastic pollution threatens the environment, health, and livelihoods of their communities. They say there is too much at risk to wait.

# SOURCE NOTES


## Introduction: Living in the Age of Plastic

1. Matt Simon, *A Poison like No Other: How Microplastics Corrupted Our Planet and Our Bodies.* Washington, DC: Island, 2022, p. 7.
2. Lisa Song, "Selling a Mirage: The Deception Behind Plastic Recycling," ProPublica, June 20, 2024. www.propublica.org.
3. Quoted in Dynahlee Padilla-Vasquez, "Protect Our Planet from Plastic Pollution: 5 Things to Know," United Nations Foundation, May 1, 2024. https://unfoundation.org.
4. Quoted in Britt Wray, "I Want to Be More Radical, but I Can't," Gen Dread, May 23, 2024. https://gendread.substack.com.

## Chapter One: The Plastics Revolution

5. A.J. Jacobs, "Trying to Live a Day Without Plastic," *New York Times*, January 11, 2023. www.nytimes.com.
6. Jacobs, "Trying to Live a Day Without Plastic."
7. Quoted in Institute of Museum and Library Services, "Bakelite: The 'Material of a Thousand Uses,'" Henry Ford, October 15, 2015. www.thehenryford.org.
8. Quoted in Jeffrey L. Meikle, *American Plastic: A Cultural History.* New Brunswick, NJ: Rutgers University Press, 1995, p. 180.
9. Susan Freinkel, *Plastic: A Toxic Love Story*. Boston: Houghton, Mifflin, Harcourt, 2011, p. 7.
10. Quoted in Chris Cillizza, "What If We Lived in a World Without Plastic?," *Downside Up*, CNN, October 31, 2022. www.cnn.com.
11. Quoted in Cillizza, "What If We Lived in a World Without Plastic?"
12. *Life*, "Throwaway Living: Disposable Items Cut Down Household Chores," August 1, 1955, p. 43.
13. Freinkel, *Plastic*, p. 10.

## Chapter Two: The Threat to the Planet, Animals, and Human Health

14. Quoted in Brevard Zoo, "Plastic Pollution Has Detrimental Effects on Young Sea Turtles," January 21, 2024. https://brevardzoo.org.
15. United Nations Environment Programme, "Visual Feature: Beat Plastic Pollution," 2022. www.unep.org.
16. Erica Cirino, *Thicker than Water: The Quest for Solutions to the Plastic Crisis.* Washington, DC: Island, 2021, p. 7.
17. Michiel Roscam Abbing, *Plastic Soup: An Atlas of Ocean Pollution*. Washington, DC: Island, 2019, p. 42.
18. Cirino, *Thicker than Water*, p. 45.
19. Quoted in Sandee LaMotte, "Which Foods Have the Most Plastics? You May Be Surprised," CNN, April 22, 2024. www.cnn.com.
20. Quoted in Simon, *A Poison like No Other*, p. 154.

## Chapter Three: The Reality of Plastics Recycling

21. Quoted in National Public Radio, *Is Recycling Worth It Anymore? The Truth Is Complicated*, YouTube, April 21, 2021. www.youtube.com/watch?v=iBGZtNJAt-M.
22. Quoted in National Public Radio, *Is Recycling Worth It Anymore?*
23. Quoted in Laura Sullivan, "Plastic Wars: Industry Spent Millions Selling Recycling—to Sell More Plastic," *All Things Considered*, National Public Radio, March 31, 2020. www.npr.org.
24. Quoted in *Frontline*, "How the Plastics Industry Used Recycling to Fend Off Bans," PBS, March 31, 2020. www.pbs.org.
25. Quoted in Sullivan, "Plastic Wars."
26. Quoted in Plastics Industry Association, "Plastics Industry Association Launches Recycling Is Real Advocacy Campaign," September 14, 2023. www.plasticsindustry.org.
27. Oliver Franklin-Wallis, *Wasteland: The Dirty Truth About What We Throw Away, Where It Goes, and Why It Matters*. London: Simon & Schuster, 2023, p. 60.
28. Quoted in Elizabeth Kolbert, "How Plastics Are Poisoning Us," *New Yorker*, June 26, 2023. www.newyorker.com.
29. Franklin-Wallis, *Wasteland*, p. 75.

30. Quoted in Rethink Plastic Alliance, "'We Are Now Plastic Farmers'—the Human Cost of Waste Colonialism," August 12, 2022. https://rethinkplasticalliance.eu.

## Chapter Four: Curbing Plastic Pollution

31. Claire Caulfield, "The Problem with Honolulu's Single-Use Plastic Ban at Restaurants," Honolulu Civil Beat, July 22, 2021. www.civilbeat.org.
32. Annie Leonard and Martin Bourque, "Op-Ed: Berkeley Isn't Just Attacking Plastic Waste, It's Rejecting our Entire Throw-away Culture," *Los Angeles Times*, January 27, 2019. www.latimes.com.
33. Quoted in Tik Root, "Inside the Long War to Protect Plastic," Center for Public Integrity, May 16, 2019. https://publicintegrity.org.
34. Faran Savitz, "New Report: Analysis Finds Bag Bans Effective at Reducing Plastic Waste, Litter," Penn Environment Research and Policy Center, January 18, 2024. https://environmentamerica.org.
35. Quoted in Adrienne Matei, "Curb 'Stupid Plastics' and Stop Industry BS: Urgent Actions to Prevent a Plastic Crisis," *The Guardian* (Manchester, UK), July 9, 2024. www.theguardian.com.
36. Sierra Club, "Support the Maryland Bottle Bill to Reduce Litter and Plastic Pollution, and Increase Recycling," AddUp. https://addup.sierraclub.org.
37. Quoted in Michael Corkery, "Beverage Companies Embrace Recycling, Until It Costs Them," *New York Times*, July 4, 2019. www.nytimes.com.
38. Annie Leonard, *The Story of Stuff: The Impact of Overconsumption on the Planet, Our Communities, and Our Health—and How We Can Make It Better.* New York: Free Press, 2010, p. 233.
39. Beyond Plastics, "Extended Producer Responsibility (EPR)." www.beyondplastics.org.
40. Quoted in Break Free from Plastic, "Plastic Treaty Talks Stall Despite Support for Production Cuts, Additional Session Planned," December 2, 2024. www.breakfreefromplastic.org.

# ORGANIZATIONS AND WEBSITES

**Algalita**
https://algalita.org
Founded by Charles Moore, who put ocean plastic on the global agenda, Algalita engages students and educators in raising awareness of plastic pollution. The site's student action guides include instructions for tracking where local plastic waste ends up, conducting a brand audit, and designing a community compost system.

**Beyond Plastics**
www.beyondplastics.org
Beyond Plastics combines policy expertise and grassroots advocacy to end plastic pollution. Its website includes a book club, articles, and fact sheets, along with reports about plastic pollution, as well as helpful advocacy tools and templates, including guidance on how to organize a school or campus plastics-free movement.

**Break Free from Plastic**
www.breakfreefromplastic.org
Break Free from Plastic is a membership-based, global organization that is pushing to find solutions to the plastic pollution crisis. The site includes reports from past brand audits and tools to organize a new brand audit, resources for young people on the zero-waste movement, and instructions for launching a plastics-free school or campus campaign.

**Keep America Beautiful**
https://kab.org
Sponsored by some of the world's biggest beer, soda, cigarette, and other plastics-producing corporations, Keep America Beautiful was founded in response to public pressure to cut waste. The organization supports a national cleanup day, a national recycling day, and community tree plantings, and it sends out its signature cleanup kits by request.

**Ocean Conservancy**
https://oceanconservancy.org
Ocean Conservancy works to protect oceans from challenges such as plastic pollution, ocean warming, and overfishing. Its website features a blog tracking ocean-related environmental issues, fun facts about marine animals, reports from its International Coastal Cleanups, and details on how to join a cleanup as a citizen scientist.

**Plastic Pollution Coalition**
www.plasticpollutioncoalition.org
This high-profile coalition includes nonprofit groups, eco-friendly businesses, and individuals, including celebrities working for a world free of plastic pollution and its toxic impacts. Its website hosts a resource library, plus articles and videos about its Youth Ambassadors program and public relations materials for its Last Plastic Straw and other campaigns.

**Sea Turtle Conservancy**
https://conserveturtles.org
The world's oldest sea turtle research and conservation group, the Sea Turtle Conservancy works to save sea turtles from extinction. Its website describes various sea turtle species and the threats posed to them by single-use plastics. Visitors to the site can download an app to track sea turtles around the world.

**Story of Stuff Project**
www.storyofstuff.org
The Story of Stuff encourages honest conversation about today's consumption-crazed culture and its environmental impact. Its website provides a series of entertaining, informative *Story of* animated video shorts about plastic, bottled water, microbeads, and other topics, as well as the feature-length documentary *The Story of Plastic*.

# FOR FURTHER RESEARCH


## Books

Erica Cirino, *Thicker than Water: The Quest for Solutions to the Plastics Crisis*. Washington, DC: Island, 2021.

Oliver Franklin-Wallis, *Wasteland: The Dirty Truth About What We Throw Away, Where It Goes, and Why It Matters*. London: Simon & Schuster, 2023.

Danielle Smith-Llera, *You Are Eating Plastic Every Day: What's in Our Food?* North Mankato, MN: Compass Point, 2019.

Imari Walker-Franklin and Jenna Jambeck, *Plastics: Essential Knowledge Series*. Cambridge, MA: MIT Press, 2023.

## Internet Sources

Basel Action Network, "Plastic Waste Trade Data," 2025. www.ban.org.

*Frontline* and National Public Radio, *Plastic Wars*, PBS, 2020. www.pbs.org.

Laura Parker, "How the Plastic Bottle Went from Miracle Container to Hated Garbage," National Geographic, August 23, 2019. www.nationalgeographic.com.

Plastics Industry Association, "Recycling Is Real Across America," Recycling Is Real, 2024. https://recyclingisreal.com.

Laura Sullivan et al., "The Myth of Plastic Recycling," *Short Wave*, National Public Radio, December 12, 2022. www.npr.org.

United Nations Environment Programme, "Visual Feature: Beat Plastic Pollution," 2022. www.unep.org.

Urban Green Council, "Virtual Tour: Sims Municipal Recycling in Brooklyn, New York with Karen Napolitano, Education and Outreach Coordinator," June 21, 2021. www.urbangreencouncil.org.

US Environmental Protection Agency, "Frequently Asked Questions About Plastic Recycling and Composting," November 21, 2024. www.epa.gov.

# INDEX

## PICTURE CREDITS

Cover: Aditya Agung Santoso/Shutterstock

6: Trong Nguyen/Shutterstock
10: Legitaa/Shutterstock
12: Meaw_stocker/Shutterstock
15: Juliya Shangarey/Shutterstock
20: David Savlatori/VWPics/Alamy Stock Photo
25: Paulo de Oliveira/NHPA/Avalon.red/Newscom
28: Sergii Rudiuk/Alamy Stock Photo
32: RICHARD B. LEVINE/Newscom
34: Carolyn jenkins/Alamy Stock Photo
40: Jahangir Alam Onuchcha/Shutterstock
44: picture alliance/Frank Duenzl/Newscom
46: Maria Kraynova/Alamy Stock Photo
52: imageBROKER.com/Alamy Stock Photo

## ABOUT THE AUTHOR

Meryl Loonin writes nonfiction books for young adult readers. She has worked in documentary film and TV and co-founded an organization to support teen library programs. She hopes this book will inspire readers to take part in community and worldwide efforts to curb plastic pollution.